STARS

Rebecca Woodbury, Ph.D., M.Ed.

Gravitas Publications Inc.

STARS

Illustrations: Janet Moneymaker

Stars
ISBN 978-1-950415-45-8

Published by Gravitas Publications Inc.
Imprint: Real Science-4-Kids
www.gravitaspublications.com
www.realscience4kids.com

RS4K

Photo credits: Cover & Title Pg: John Badham, CC BY SA 4.0; Above: NASA Goddard; P.5. ESA/Hubble & NASA, R. Cohen-CC BY 2.0; P.9. Nova Dawn Astrophotography, CC BY SA 4.0; P.11. Skatebiker, CC By SA 3.0; P.13. Mellostorm, CC BY SA 3.0; P.19. Top, NASA/JPL; Bottom, NASA

The **Sun** is a **star.** It heats
our Earth and gives us light.

We love
the Sun!

Yes!

Our Sun is not the only star that exists. There are even more stars than the ones we can see in the sky.

I can't count that high.

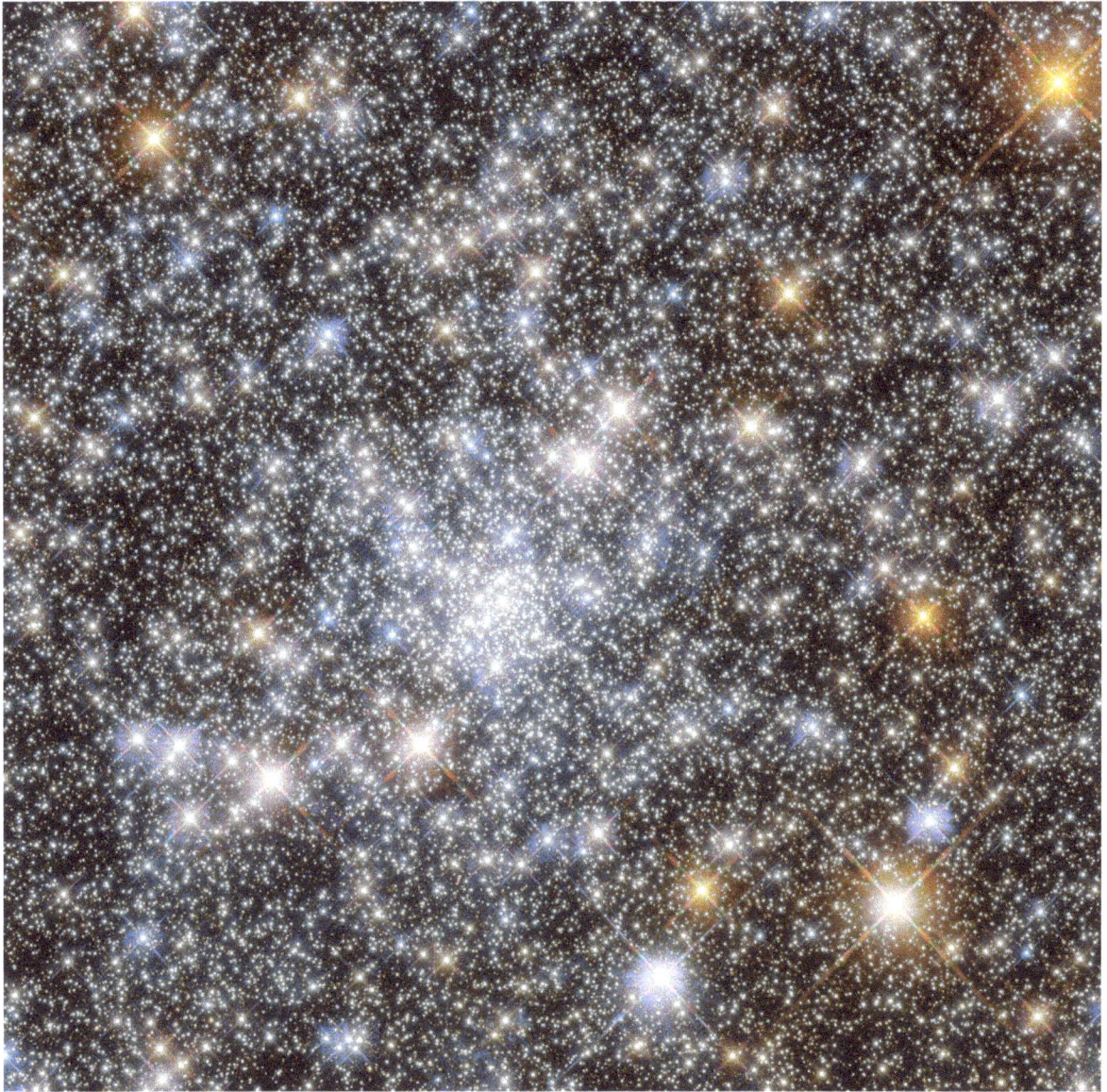

A **star** is an object in space that is made of **hydrogen** and **helium atoms**. It makes its own **light energy and heat energy** when hydrogen atoms combine to make helium.

Review: ATOMS

- **Atoms** are tiny building blocks that can link together.

- **Atoms** make up everything we touch, taste, smell, and see.

- **Matter** is the name for everything we touch, taste, smell, and see.

Stars are different sizes.

Some are smaller than our Sun.

Some are much bigger.

The stars nearest to our Sun are **Proxima Centauri, Alpha Centauri A,** and **Alpha Centauri B.**

Alpha Centauri A is a little bigger than the Sun. Alpha Centauri B is slightly smaller, and Proxima Centauri is a lot smaller.

Can you see Proxima Centauri?

I think it's too small.

- Alpha Centauri A is the bright star on the left.
- Alpha Centauri B is the bright star on the right.
- Proxima Centauri is inside the red circle.

Sirius is the brightest star we see in the sky.

Sirius is seriously bright!

Proxima Centauri is closer to us than Sirius is. But Sirius looks closer because it is much bigger than Proxima Centauri and makes so much more light.

WHY DOES THAT STAR LOOK BRIGHTER THAN US? IT'S FARTHER AWAY!
I MAKE MORE LIGHT, SO I LOOK CLOSER!

The biggest star we can see is **VY Canis Majoris.** It is called **VY CMa** for short.

VY CMa is about 2,000 times bigger than the Sun. If our Sun were as big as VY CMa, most of the planets would be inside it.

Astronomers think that most stars
have **planets** orbiting them.

Orbiting?

Orbiting means
moving in a path
around an object.

Illustrations of what other planets and suns might look like.

Astronomers look deep into space hoping to find another planet like Earth that has life on it.

What do you think life on another planet would look like?

Good question.

How to say science words

Alpha Centauri A (AAL-fuh sen-TAWR-ee AY)

Alpha Centauri B (AAL-fuh sen-TAWR-ee BEE)

astronomer (uh-STRAH-nuh-mer)

atom (AA-tum)

energy (EN-uhr-jee)

heat (HEET)

helium (HEE-lee-uhm)

hydrogen (HIY-druh-juhn)

light (LIYT)

matter (MAA-tuhr)

orbit (AWR-buht)

planet (PLAA-nuht)

Proxima Centauri (PRAHK-suh-muh sen-TAWR-ee)

science (SIY-uhns)

Sirius (SIR-ee-uhs)

star (STAHR)

VY Canis Majoris (VEE WIY KAY-nuhs muh-JAWR-uhs)

VY CMa (VEE WIY SEE EM AY)

www.ingramcontent.com/pod-product-compliance
Lightning Source LLC
Chambersburg PA
CBHW040152200326
41520CB00028B/7580